Drip Irrigation's Delicate Dance

Precise Irrigation Techniques For Optimizing Water Utilization And Enhancing Agricultural Yields

Lydia Howard

Contents

1 Introduction to Drip Irrigation

Drip irrigation, unlike flood or furrow irrigation, delivers water directly to the plant roots. This targeted approach minimizes water waste from runoff and evaporation, a significant advantage in arid or semi-arid regions where water resources are scarce. The precision of drip irrigation allows for a more efficient use of water, ultimately leading to higher crop yields and reduced operational costs. Consequently, farmers can see a substantial return on their investment, despite the initial setup expenses. It's this careful balancing of initial investment and long-term savings that makes drip irrigation increasingly attractive.

The advantages extend beyond just water savings. Because water is applied directly to the root zone, weeds have less access to moisture, thus reducing competition for resources and minimizing the need for herbicides. Furthermore, the consistent moisture at the root level promotes healthier plant growth, leading to improved fruit and vegetable quality and increased yields. This improved quality can often translate to higher market prices and increased profitability for farmers. The consistent moisture also supports better nutrient uptake, maximizing the benefit of fertilizers.

Consider the labor aspect. Drip irrigation systems require less manual labor compared to traditional methods. Time spent filling ditches, leveling fields, and manually watering plants is significantly reduced, freeing up valuable time for other farm operations. This reduced labor requirement directly translates into cost savings, improving farm profitability and allowing farmers to focus on other essential tasks such as pest management and crop monitoring. Efficient resource allocation is key to successful farming, and drip irrigation facilitates this efficiency.

Effective system design is paramount for successful implementation. A thorough site assessment is crucial; this involves analyzing factors like soil type, topography, and the specific needs of the crop to be irrigated. Proper soil analysis helps determine the appropriate emitter spacing and flow rates to ensure uniform water distribution throughout the field. The selected emitter type should also perfectly complement the soil texture and crop root system for optimal performance. The topography influences the design, necessitating adjustments for slopes and uneven terrains to maintain uniform water pressure throughout the entire system.

Choosing the correct water source and implementing adequate filtration is another critical step. The water source's quality directly impacts the system's longevity and efficiency. Sediment, dissolved minerals, and other contaminants can clog emitters, reducing the system's performance and requiring more frequent maintenance. A suitable filtration system, carefully chosen to handle the specific contaminants present in the water source, is thus essential to prevent such problems. Regular maintenance of the filters is key to continuous efficient water flow.

Once the water source and filtration are secured, the system layout needs careful planning. This involves designing an efficient network of pipes and tubing to distribute water evenly to all plants. The layout should be optimized to minimize pipe length, reducing costs and pressure loss. Careful consideration must be given to the placement of emitters to ensure consistent water delivery to each plant's root zone. A well-designed layout will reduce inconsistencies in growth across the field and result in improved yields.

Finally, ongoing monitoring and maintenance are essential for the long-term success of any drip irrigation system. Regularly inspecting the system for leaks, clogged emitters, and other problems allows for prompt repairs, preventing significant water loss and crop damage. A proactive maintenance schedule, including cleaning and replacing worn-out components, ensures the system's continued effectiveness and efficiency. Routine checks and timely adjustments will minimize downtime and maximize the system's lifespan, providing a long-term return on the initial investment. The long-term benefits far outweigh the initial investment, making drip irrigation a sound choice.

1.1 What is Drip Irrigation?

Drip irrigation is a localized irrigation system that delivers water directly to the roots of plants, minimizing water waste and maximizing efficiency. Unlike traditional methods like flood or furrow irrigation, which often lead to significant water loss through evaporation and runoff, drip irrigation provides targeted hydration, focusing water delivery precisely where it's needed most. This targeted approach conserves water resources, proving particularly crucial in arid or semi-arid regions where water scarcity is a pressing concern. It allows for precise control over water application, ensuring that the plant receives the ideal amount of moisture, leading to healthier growth and higher yields. The system uses a network of pipes, tubing, and emitters to deliver water slowly and steadily, mimicking the natural rainfall process but with greater control. Such precision minimizes water stress and promotes optimal root development, ultimately leading to improved plant health and productivity.

The system's design comprises several key components. A water source, often a well or reservoir, provides the raw water. This is then typically filtered to remove sediments and other impurities that could clog the emitters and compromise the system's functionality. The filtered water then travels through a network of larger diameter pipes, which branch into smaller pipes and finally into the individual emitters. These emitters are carefully positioned near the base of the plant, allowing for a slow and steady release of water directly into the root zone. The spacing and placement of emitters are critically important, ensuring even water distribution and preventing uneven growth. This meticulous placement demands careful planning and accurate execution, a factor which contributes to the system's overall efficiency.

Choosing the right emitters is crucial for optimal system performance. Emitters come in a variety of designs and flow rates, each suited to specific plant types and soil conditions. Selecting the wrong emitter can significantly impact the efficiency of the system; undersized emitters may not deliver sufficient water, whereas oversized emitters can lead to excessive water use and potential runoff. In addition to emitter selection, the type and size of tubing used to carry the water are equally important. The tubing must be durable enough to withstand the pressure of the system and the environmental conditions, especially in areas prone to harsh weather such as intense sun exposure and extreme temperature fluctuations. The choice of tubing material, often polyethylene, should be appropriate for the application, considering factors like soil type, soil temperatures and system pressure. Appropriate connectors, both for joining pipes and securing emitters, also need careful selection to ensure water tightness and system reliability. Effective irrigation scheduling depends on several crucial factors. Soil moisture levels are paramount, with appropriate monitoring techniques such as tensiometers or soil moisture sensors used to determine when irrigation is necessary. These measurements, alongside weather data including rainfall, temperature, and wind speed, provide valuable information for scheduling irrigation events. Additionally, the growth stage of the crop significantly influences irrigation needs, with young seedlings generally requiring less water than mature plants. Soil type also plays a critical role, with sandy soils requiring more frequent irrigation than heavier clay soils due to their improved drainage capabilities. This requires irrigation managers to adjust watering schedules accordingly, ensuring that water is applied efficiently and effectively to meet the plant's requirements throughout its entire lifecycle.

Water quality is a critical aspect often overlooked. The presence of certain salts, minerals or biological matter can cause clogging of emitters, reducing the overall effectiveness of the system. This emphasizes the need for pre-filtration before the water enters the system. Regular monitoring of water quality, alongside proper filter maintenance, helps ensure the continued smooth operation of the irrigation system. Moreover, the system's ability to deliver fertilizers directly to the roots via fertigation - the simultaneous application of fertilizers and water - can drastically improve nutrient management. Applying fertilizers precisely to the plant's root zone reduces wastage, ensures optimal nutrient uptake and minimizes environmental impact. This method also allows for highly precise control over nutrient application, adjusting the nutrient solution to meet the plant's needs at different growth stages, which contributes significantly to the plant's overall health and nutritional balance. Understanding the potential impact of water chemistry on both the irrigation system and plant health is a crucial aspect of successful drip irrigation.

Regular maintenance is essential for longevity and optimum system performance. One common problem is emitter clogging, often caused by sediment buildup or biological growth within the emitters. Regular flushing of the system can help prevent this, along with appropriate water filtration. Leaks in the system, while less common, should be addressed promptly to minimize water loss and avoid potential damage. Inspecting the entire system regularly for any sign of damage or malfunction, including checking the integrity of all connections, valves, and emitters, is vital. A well-maintained system will result in optimal water usage, minimizing waste and maximizing crop yield. This should include periodic cleaning of filters, replacement of worn-out components, and thorough flushing of the entire network to remove any accumulated debris or minerals.

In conclusion, understanding the intricate details of drip irrigation is paramount to utilizing it effectively. From careful system design and emitter selection to meticulous scheduling and diligent maintenance, each aspect plays a vital role in maximizing water use efficiency and optimizing crop yields. The integration of technologies that enable precise water management offers further improvement towards achieving sustainable agriculture. By carefully considering all these factors, growers can harness the power of drip irrigation to produce healthier, more productive crops while conserving precious water resources.

1.2 Advantages of Drip Irrigation

Reduced water consumption is a significant advantage. Drip irrigation delivers water directly to the plant roots, minimizing evaporation and runoff, unlike traditional methods which often lose a substantial portion of water to these processes. This targeted approach conserves precious water resources, especially crucial in arid and semi-arid regions where water scarcity is a major constraint. The efficiency gains translate to lower water bills and a reduced environmental footprint, making drip irrigation a sustainable choice for farmers and the environment alike. Furthermore, this precision allows for optimized water use, maximizing the impact of every drop.

Improved water use efficiency directly translates to higher yields. By providing water precisely where it's needed, drip irrigation ensures consistent soil moisture, promoting optimal plant growth and development. This contrasts sharply with flood irrigation or sprinkler systems, which often lead to uneven moisture distribution, resulting in stunted growth and reduced yields. The consistent moisture supply from drip irrigation enhances nutrient uptake, further boosting productivity and contributing to healthier, more vigorous plants, ultimately maximizing the return on investment for farmers. This consistent moisture level is crucial for optimal crop development.

Reduced weed growth is another significant benefit. Drip irrigation's localized water application restricts moisture availability to areas surrounding the plants, hindering weed germination and growth. This reduces the need for herbicides, leading to lower costs, reduced environmental impact, and improved crop quality. The absence of widespread wetting also helps prevent the spread of soil-borne diseases, enhancing crop health and minimizing the need for pesticides. This contributes to sustainable and environmentally friendly agricultural practices.

Lower labor costs are a major economic advantage. Traditional irrigation methods often require significant manual labor for tasks such as leveling fields, constructing and maintaining canals, and managing water distribution. Drip irrigation systems, on the other hand, require minimal manual intervention once they're established. This translates to reduced labor costs and allows farmers to focus their time and resources on other critical aspects of farming, improving overall farm efficiency and profitability. The time saved can be directed to other value-added tasks.

Increased profitability is a key outcome. The combined effects of increased yields, reduced water and labor costs, and lower herbicide use contribute to a significant boost in overall farm profitability. Drip irrigation enables farmers to produce higher-quality crops while minimizing input costs, making it an economically viable and sustainable solution for maximizing profits. The improved efficiency and reduced resource waste contribute to a more stable and resilient agricultural operation, ensuring long-term economic viability.

Enhanced soil health is a long-term benefit often overlooked. The localized water application prevents soil erosion and compaction, common problems associated with surface irrigation methods. The improved soil structure encourages better aeration and drainage, enhancing root penetration and nutrient availability. This creates a healthier and more productive soil environment, sustaining long-term crop yields and reducing the need for soil amendments. A healthier soil structure forms the bedrock of sustainable agriculture.

Better control over fertilization is another advantage. Fertigation, the process of applying fertilizers through the drip irrigation system, offers precise nutrient application directly to the plant roots. This increases nutrient uptake efficiency, reducing fertilizer waste and minimizing environmental pollution associated with excessive fertilizer application. The targeted approach minimizes nutrient runoff into water bodies, protecting water quality and reducing the overall environmental impact of fertilization practices. Precise control of nutrient application is essential for environmentally friendly farming.

Improved crop quality is an important factor in modern agriculture. The consistent moisture and nutrient supply provided by drip irrigation leads to improved crop quality, resulting in larger, healthier fruits and vegetables with enhanced taste and appearance. This improved quality commands higher market prices, translating to increased revenue for farmers and a better product for consumers. Drip irrigation contributes to producing high-quality agricultural goods.

Greater adaptability to diverse terrains and crops is a remarkable asset.

Unlike some irrigation methods limited to specific topographic conditions, drip irrigation is adaptable to various terrains, from flat fields to sloped lands. Its versatility extends to a wide range of crops, from field crops to orchards and high-value horticultural crops, making it a truly versatile irrigation solution for diverse farming systems. This adaptability ensures its effectiveness across varied agricultural landscapes. Reduced pest and disease issues is a less apparent but crucial advantage. The localized water application limits the spread of many soil-borne diseases and pests, minimizing the risk of crop damage and the need for chemical interventions. The controlled environment also helps to reduce the humidity around the plants, creating a less favorable environment for certain fungal diseases. This holistic approach contributes to a healthier and more sustainable agricultural system. The reduced reliance on pesticides translates to lower costs and a healthier environment.

2 System Design & Planning

Careful site assessment is paramount. Begin by thoroughly analyzing your land's topography, noting slopes and elevations to ensure even water distribution. Understanding your soil type—its texture, drainage, and water-holding capacity—is equally vital. This dictates emitter selection and spacing, influencing the frequency and duration of irrigation. Finally, consider the specific water requirements of your chosen crops at different growth stages. This detailed understanding will form the bedrock of your design, preventing costly mistakes and maximizing efficiency from the outset.

A reliable water source is essential. This could be a well, a reservoir, or municipal supply. Regardless of the source, water quality is crucial. High sediment levels or excessive mineral content can clog emitters, reducing efficiency and potentially damaging the system. Therefore, appropriate filtration is mandatory. This may involve sand filters, screen filters, or even more sophisticated systems depending on your water's quality. Regular maintenance of your filtration system is also key; neglecting this can quickly negate your efforts. Invest in a system appropriate for your scale of operation and the quality of your water.

The system's layout needs careful planning. Consider the field's shape and size, positioning the main pipeline strategically to minimize pipe length and pressure loss. Sub-mains should branch out from this main line, feeding laterals that deliver water directly to the emitters. Remember, uniform water distribution is the ultimate goal. To achieve this, ensure appropriate spacing between emitters based on your soil type and crop water needs. Poor planning can lead to uneven water distribution, resulting in poor crop growth and wasted resources. Careful consideration of your layout will pay dividends in the long run. Use a detailed drawing to map out your design before purchasing equipment.

Choosing the correct emitters is crucial. Emitters come in various types, each with a unique flow rate and pressure requirement. Selecting the appropriate emitter depends heavily on your soil type, crop water requirements, and the overall system pressure. For example, clay soils may require low-flow emitters to prevent runoff, whereas sandy soils might need higher-flow emitters to compensate for rapid drainage. Carefully matching the emitter type to the specific conditions of each area of your field will ensure optimal water delivery and prevent water loss.

Tubing and connectors play a critical role in system longevity and performance. The material selected should be durable enough to withstand the pressure and environmental conditions. Polyethylene tubing is a common choice due to its flexibility and resistance to corrosion. Ensure the connections are secure and leak-proof to prevent water loss and maintain consistent pressure throughout the system. The selection of high-quality connectors is not just about initial cost savings, but also about minimizing future repairs and maximizing the system's lifespan. Invest wisely here.

Pressure regulation is essential for uniform water distribution. Pressure variations within the system can cause uneven water delivery, leading to inconsistent crop growth. Pressure-regulating valves are crucial for managing pressure across the entire system, particularly in uneven terrains. They ensure that water reaches all emitters at the intended flow rate. Without this, your investment in precision irrigation will be significantly undermined. Regular checks on the pressure are vital to maintaining consistent performance.

2.1 Site Assessment

Successful drip irrigation hinges on a meticulous site assessment. Begin by thoroughly evaluating your land's topography. Steep slopes necessitate adjustments in layout and potentially the use of pressure-compensating emitters to prevent uneven water distribution. Gentle slopes present fewer challenges, but careful planning remains crucial to ensure efficient water flow throughout the system. Consider the overall land area and divide it into manageable zones, each with its own irrigation sub-unit for better control and easier maintenance. This segmented approach also allows for customized irrigation based on varied soil conditions within the overall site.

Soil analysis is paramount. Determine your soil's texture—whether sandy, silty, clayey, or a combination—as this directly impacts water infiltration and retention rates. Sandy soils drain rapidly, requiring more frequent, albeit shorter, irrigation cycles. Clayey soils retain water for extended periods, necessitating less frequent, yet longer, watering sessions. A professional soil test will reveal crucial information regarding nutrient content, pH level, and potential salinity issues—factors that significantly influence both crop health and the effectiveness of fertigation. Understanding these parameters allows you to optimize irrigation scheduling and nutrient application, maximizing yield while minimizing waste. Don't overlook the impact of subsurface drainage. Poor drainage can lead to waterlogging, hindering root development and potentially damaging crops.

Next, consider the prevailing climatic conditions. Local rainfall patterns significantly influence irrigation scheduling. Regularly monitor temperature, humidity, and solar radiation levels to assess evapotranspiration rates. Areas experiencing high temperatures and intense sunlight will necessitate increased irrigation frequency compared to cooler, shadier locations. Wind speed also plays a role, influencing evaporation rates and potentially affecting emitter placement to reduce water loss. Accurate weather forecasting is invaluable in predicting irrigation needs and adjusting your irrigation schedule proactively. This predictive approach helps maximize water efficiency and minimizes environmental impact. You also need to understand the wind patterns in your area. Strong, consistent winds can significantly affect the distribution of water from the emitters and create uneven watering across the field, resulting in uneven growth and reduced yields. Consider using windbreaks or adjusting emitter placement to mitigate this issue.

The chosen crop dictates specific requirements. Different plants have varied water needs depending on their growth stage, root systems, and overall water requirements. Young seedlings generally require less water than mature plants. Deep-rooted crops tolerate drier conditions compared to shallow-rooted species. Consult appropriate agricultural resources or consult with an experienced agronomist to determine the precise water requirements for your specific crops. This will ensure that your irrigation system provides the optimum amount of water at each growth stage, avoiding both overwatering and underwatering which can both impact yield and fruit quality. Knowing this information lets you plan the correct emitter flow rate and spacing for maximum effect.

Finally, access to a reliable water source is essential. This includes assessing both water quantity and quality. Sufficient water volume must be available to meet the irrigation demands throughout the growing season, considering peak water usage periods. Water quality is crucial to prevent emitter clogging. High levels of sediment, salts, or organic matter necessitate filtration systems to protect your drip irrigation components. Regular water quality testing ensures that the water used is suitable for both the plants and the irrigation system. Having this information will allow you to tailor your system to prevent damage and ensure effective irrigation. Consider proximity to the water source as well. The distance between the water source and the irrigation area impacts the system's overall design and the necessary pumping capacity. Longer distances often mean higher energy costs. Optimizing this factor will contribute to the overall cost-effectiveness of the project.

2.2 Water Source & Filtration

A reliable water source is the lifeblood of any successful drip irrigation system. Consider your options carefully; municipal water supplies offer consistency but may incur higher costs. Surface water sources like ponds or rivers can be free, but their quality fluctuates with rainfall and require careful monitoring for contaminants. Groundwater, accessed through wells, offers a consistent supply, although initial investment can be significant, and water quality needs assessment before implementation. Each choice presents unique advantages and disadvantages regarding cost, accessibility, and water quality. The selection should align with your budget, available resources, and long-term sustainability goals. Regular testing of your chosen water source is crucial to maintain the system's health.

Water quality directly impacts the longevity and efficiency of your drip irrigation system. High levels of sediment, organic matter, or dissolved minerals can swiftly clog emitters, reducing water flow and ultimately harming crop yield. Therefore, a robust filtration system is non-negotiable. A multi-stage approach is generally recommended, starting with a coarse screen filter to remove larger debris. This initial filtration prevents larger particles from damaging the finer filters downstream. Follow this with a sand filter for finer particulate removal. Depending on your water source's specific challenges, you might also integrate a disc filter or a pressure filter for even more refined filtration. Consider adding a final filter specifically designed for drip irrigation systems to remove the tiniest particles that could cause clogging. Regular cleaning and maintenance of these filters are paramount to sustained performance.

The choice of filter type will significantly influence your system's performance and maintenance schedule. Sand filters are cost-effective and relatively easy to maintain. However, they may require more frequent backwashing compared to other options. Disc filters offer a higher degree of filtration, effectively removing even smaller particles. Yet, they often require more specialized cleaning procedures and might be more expensive to acquire. Pressure filters are compact and effective but may demand more frequent cartridge replacements depending on your water quality. A thorough understanding of your water's composition and the specific capabilities of each filter type will guide you to the optimal selection for your system.

Remember, the filtration system isn't a standalone element; its effectiveness depends heavily on its placement within the overall irrigation design. Place filters strategically upstream of the pump, ensuring all water passes through the purification process before reaching the emitters. Locate them easily accessible for routine maintenance and cleaning, to minimize downtime. Adequate space around the filter is equally important to allow for convenient access and to prevent damage during these procedures. The chosen filter needs to be compatible with the flow rate and pressure of your entire drip irrigation network. Selecting a filter that's too small will restrict flow, and selecting one that's too large may lead to insufficient filtration. Accurate sizing ensures optimal system efficiency and prolongs the lifespan of your emitters.

Regular maintenance isn't just about extending the system's lifespan; it's fundamental for maintaining consistent water delivery and maximizing crop yield. Regular inspection of your filtration system, which might entail a visual check for any damage or leaks, is crucial. Regular cleaning schedules should be established based on your water quality and filter type. Backwashing a sand filter, for instance, might be required weekly depending on conditions. Disassembling and cleaning disc filters or replacing cartridges in pressure filters should follow the manufacturer's guidelines. Proper documentation of these cleaning intervals and any maintenance actions will aid you in identifying potential issues and optimising your cleaning schedule. By meticulously tending to your filtration system, you ensure optimal performance of the entire drip irrigation network.

The integration of a well-maintained filtration system within your drip irrigation setup translates into more than just preventing clogging. It ensures the uniform distribution of water to each plant, leading to healthier plants and ultimately higher yields. Furthermore, a clean filtration system prolongs the life of your emitters, saving you costs associated with frequent replacements. Moreover, cleaner water helps maintain the system's efficiency, decreasing energy costs associated with pumping. Investing time and effort into water source selection and filtration pays off exponentially in improved crop production, cost savings and a more sustainable irrigation practice. By prioritizing water quality, you ensure the long-term success and sustainability of your drip irrigation system.

2.3 Layout & Design Principles

Careful planning is paramount to a successful drip irrigation system. Begin by creating a detailed site map, noting the topography, soil type, and the location of existing infrastructure like fences or buildings. This map will serve as your blueprint, guiding the placement of laterals and the main water supply line. Consider the slope of your land; steeper slopes may require more frequent support structures for your tubing to prevent sagging and uneven water distribution. Remember that accurate measurements are crucial to avoid costly mistakes down the line, ensuring that your design perfectly matches the area you need to irrigate. This initial planning phase will save you considerable time and resources in the long run.

Next, meticulously map out the layout of your drip irrigation network. This involves determining the optimal placement of the main lines, laterals, and emitters based on your site map and the specific needs of your crops. Consider using computer-aided design (CAD) software for more complex layouts, facilitating precise planning and visualizing your entire system. Remember that uniformity of water distribution is key; uneven watering can lead to inconsistent crop growth and reduced yields. Therefore, thoughtful spacing of emitters is essential, and you might need to adjust the spacing based on factors such as soil type, crop density, and specific water requirements of your plants. Careful planning at this stage can ensure an even distribution of water, maximizing the efficiency of your drip system.

The choice of tubing is crucial to the overall success of your project. Consider the diameter of the tubing according to the pressure and flow rate of your system; larger diameter pipes will handle higher flow rates with less pressure loss, but at an increased installation cost. Choose durable materials capable of withstanding the environmental conditions in your area. Consider factors such as UV resistance, the potential for clogging and the long-term maintenance requirements. Think about flexibility as well; certain materials offer flexibility ideal for following contours, avoiding unnecessary bends that can restrict flow. Selecting the appropriate tubing is about finding a balance between cost, durability, and the specific needs of your irrigation setup.

Emitter selection directly impacts the effectiveness of your drip irrigation system. Each emitter should be chosen based on the water requirements of your specific crops, soil type, and the overall pressure within your system. Understand flow rates and pressure compensation features; pressure-compensating emitters are particularly helpful on uneven terrain, maintaining consistent flow even with varying pressure. Low-flow emitters may be suitable for smaller plants or those requiring less water, while higher-flow emitters are appropriate for larger, more thirsty crops. Research and select emitters that match the plant requirements and offer appropriate pressure compensation if needed. Consider the long-term aspects of emitter selection by considering their longevity and the ease of maintenance and potential for clogging.

Pressure regulation is often overlooked, yet it's absolutely critical for a well-functioning drip system. Consistent pressure ensures that all emitters deliver water at the intended rate. Without proper pressure regulation, some parts of your field may receive too much water, while others receive too little. Consider pressure regulating valves at strategic points within your system, particularly before branching off into lateral lines or sections of your field with varying elevations. This ensures even distribution across all emitters, maximizing efficiency and preventing damage caused by high water pressure. Choose regulators compatible with your system and its overall pressure requirements.

Finally, think about the long-term maintenance of your system. Regular inspection is key to early detection and prevention of problems like clogged emitters or leaks. Implement a proactive maintenance schedule, including flushing the lines periodically to remove sediment and debris. Proper maintenance not only prolongs the lifespan of your drip irrigation system but also ensures consistent water delivery and optimal crop production. Consider keeping a detailed maintenance log recording dates of flushing, inspections, and any necessary repairs, thus making future maintenance simpler and more efficient. A well-maintained system will deliver a better return on your investment, providing years of efficient and reliable irrigation.

3 Emitter Selection & Placement

Choosing the right emitters is paramount; it directly impacts your system's efficiency and your crop's health. Consider the crop's water requirements, soil type, and the pressure within your system. Pressure-compensating emitters maintain a consistent flow rate even with variations in pressure along the lateral lines, a crucial feature for uneven terrain. Non-pressure-compensating emitters, while simpler and cheaper, are best suited for relatively flat areas with consistent pressure. Careful consideration of these factors ensures optimal water delivery to each plant.

Precise emitter placement is just as important as emitter selection. Uniform spacing is essential for even water distribution. The recommended spacing depends on several factors including the emitter's flow rate, the plant's water needs, and the soil type. For instance, thirsty crops in sandy soils might require closer spacing than drought-tolerant crops in clay soils. Accurate placement also minimizes water runoff and maximizes infiltration, optimizing water use efficiency. You can use different spacing patterns like parallel, staggered, or circular, depending on the crop layout and field topography. Remember to consider the plant's root zone and place the emitters accordingly.

Proper emitter installation is key to a long-lasting and efficient system. Begin by carefully preparing the land, ensuring it's free of rocks and debris that might damage the tubing or emitters. Use appropriate tools to dig trenches or create surface layouts that are conducive for your selected emitter installation method. The depth of the emitter placement depends on soil type and climate conditions. Shallow placement may lead to evaporation losses, while excessively deep placement could compromise water accessibility to roots. Always consult relevant research or local extension services for region-specific guidance to further refine installation parameters and to optimize yield and water use.

Maintaining adequate pressure is crucial for consistent emitter performance. Pressure variations can lead to uneven water distribution and reduced efficiency. Pressure-regulating valves are essential, particularly in systems with significant elevation changes or long lateral lines. These valves help maintain the ideal pressure for your chosen emitters, ensuring optimal water delivery. Install pressure gauges at strategic points to monitor pressure levels and identify potential issues promptly. Consistent monitoring is vital to keep your pressure within the specified ranges for optimal performance and longevity of your equipment.

The choice of tubing and connectors influences the system's durability and longevity. Select tubing material suitable for your specific climate and water quality. Consider factors like UV resistance, flexibility, and chemical compatibility with your chosen fertilizers or other chemicals. Choose high-quality connectors to prevent leaks, which could lead to water waste and system failure. Regularly inspect the tubing and connectors for signs of wear and tear, replacing damaged components promptly. Using quality materials will reduce long-term maintenance needs and maintain the integrity and efficiency of the system for many years to come. It's a small upfront investment that pays dividends over time.

Proper emitter selection and placement are the foundation of a successful drip irrigation system. By carefully considering emitter type, spacing, installation, pressure regulation and tubing choices you maximize water-use efficiency and enhance crop yields. Remember that local conditions will have an effect on this, and adaptation is key to the success of this project. Thorough planning and attention to detail at this stage significantly influence the overall performance and long-term success of your investment. Remember that consistent monitoring and timely maintenance are key to maximizing the benefits of drip irrigation.

3.1 Emitter Types & Flow Rates

Choosing the right emitter type is crucial for efficient drip irrigation. Consider your specific crop needs, soil type, and water pressure. For example, pressure-compensating emitters maintain a consistent flow rate even with variations in pressure across the system, ideal for uneven terrain. On the other hand, pressure-regulating emitters adjust the flow based on the system's pressure, often a more economical solution for flat areas with consistent pressure. Understanding the differences is paramount to effective system design. The wrong emitter choice could lead to uneven water distribution, impacting crop yield and potentially wasting valuable water resources. Careful planning and consideration are vital for success.

Flow rates, measured in liters per hour (L/h) or gallons per hour (GPH), directly influence the amount of water each plant receives. This needs to match the crop's water requirements, varying significantly based on growth stage, climate, and soil type. Overwatering can lead to root rot and nutrient leaching, while underwatering stunts growth and reduces yields. Precise flow rate selection requires careful assessment of various factors—soil type's water-holding capacity being particularly important. This involves understanding the rate at which the soil can absorb water and how much the plants will need for optimal performance. Matching the emitter flow rate to the soil's water intake rate ensures efficient irrigation without over-saturating the soil, a crucial aspect of sustainable water management in agriculture.

Emitter selection also impacts spacing. High flow rate emitters require greater spacing to prevent runoff and ensure even distribution. Conversely, low flow rate emitters need closer spacing to provide adequate water to plants. This intricate balance ensures every plant receives the necessary hydration without wastage. Improper spacing, regardless of emitter type, can lead to inconsistent growth and reduced yields, rendering the entire drip irrigation system inefficient, negating the benefits of targeted watering. The relationship between emitter flow rate and spacing is thus a critical aspect of system design, impacting both efficiency and yield. This is why careful calculation and strategic planning is essential.

Several emitter types exist, each with unique characteristics. In-line emitters are simple and inexpensive, suitable for straightforward systems. Their simplicity, however, can be a drawback with uneven pressure distribution or clogging. On the other hand, point-source emitters deliver water directly to the plant's root zone, minimizing water loss through evaporation and surface runoff, thus maximizing efficiency. Membrane emitters provide more consistent flow over time, minimizing clogging issues, but often at a higher cost. Choosing the right emitter type involves weighing factors such as cost, maintenance needs, and the long-term impact on system efficiency and the health of the plants. An informed decision is key to sustainable agricultural water management.

Proper emitter placement is as important as the choice of emitter itself. Emitters should be positioned close enough to the roots to deliver water efficiently. However, improper placement can lead to waterlogging or insufficient watering, depending on factors like soil type and plant root systems. Consider the root structure of the crop for optimized placement. For example, plants with extensive root systems may benefit from emitters positioned further apart, while those with shallow roots may require more closely spaced emitters. Precision in placement is crucial for obtaining maximum benefit from drip irrigation, and neglecting this could lead to inconsistencies in water delivery impacting growth and yield. This requires familiarity with the specific plant and the soil properties of the agricultural environment.

Moreover, consider factors beyond the immediate plant needs. Soil texture and structure influence the water penetration rate and runoff potential, dictating emitter type and spacing. Sandy soils drain quickly, requiring more frequent irrigation with potentially higher flow rates or more densely placed emitters. Conversely, clay soils retain water better, allowing for less frequent irrigation with lower flow rates and wider emitter spacing. Ignoring these soil-specific considerations can lead to inefficient water usage, suboptimal plant growth, and even potential soil degradation. Therefore, understanding soil conditions and plant properties is crucial for system design and operation. This careful analysis improves yields and protects soil quality.

Finally, remember that emitter selection isn't a one-time decision. Regular monitoring and maintenance are essential to ensure consistent performance. Over time, emitters can become clogged by sediment or debris, reducing their flow rate and impacting water distribution. Regular cleaning and system checks help prevent clogs and maintain the efficiency of the entire system. Proactive maintenance minimizes disruptions and optimizes the system's longevity, ensuring a continuous supply of water to the crops and preventing issues that could cause long-term problems, resulting in reduced yields and wasted resources. This ongoing effort represents a vital component of sustainable water management practices in drip irrigation.

3.2 Spacing & Distribution

Uniform water distribution is paramount for successful drip irrigation. Uneven watering leads to stressed plants in some areas and potential overwatering in others, reducing yields and wasting water. Careful planning and execution are essential to achieve consistent moisture levels across your field. This requires understanding your specific site conditions, choosing the right emitter type, and implementing precise spacing strategies.

Consider your soil type. Sandy soils drain quickly, requiring more frequent irrigation with smaller amounts of water delivered over shorter periods. Clay soils retain water longer, necessitating less frequent but potentially higher-volume watering. The emitter flow rate and spacing must be tailored to your soil's water-holding capacity. Remember, you're aiming for consistent soil moisture, not saturation. Overwatering can lead to root rot and other issues detrimental to your crops. Underwatering, naturally, leads to wilting, stunted growth, and reduced yields.

Emitter spacing directly impacts water distribution. Closer spacing provides more frequent but smaller amounts of water, ideal for sandy soils or thirsty plants. Wider spacing delivers less frequent but larger amounts of water, better suited for clay soils and plants with deeper root systems. A uniform distribution pattern will often require adjustments in emitter spacing across the field. Consider using a combination of spacings to accommodate variations in soil texture or terrain across your field. Using a standardized spacing across an uneven field might lead to uneven water application, which would contradict the primary objective of drip irrigation—optimizing water utilization and maximizing crop yields.

The optimal emitter spacing also depends on your crop's water requirements. Water-intensive plants like lettuce need more frequent watering and closer emitter spacing compared to drought-tolerant crops like certain varieties of beans. Research your target crop's specific water needs through relevant academic research or your local agricultural extension office. They may provide recommendations based on regional climates and soils. You'll then be able to accurately estimate your irrigation requirements and calibrate the system for the most efficient operation.

Remember to account for microclimates within your field. Areas exposed to full sun might dry out faster than those in shade, requiring adjusted emitter spacing or even separate irrigation zones. This fine-tuning will optimize water usage and ensure consistent watering across areas with differing environmental conditions. Even seemingly minor variations in slope can drastically affect water distribution, causing runoff in steeper areas and water pooling in low-lying sections.

Beyond emitter spacing, the layout of your mainlines and laterals significantly influences water distribution. A well-designed network ensures uniform pressure throughout the system, preventing pressure imbalances that lead to uneven watering. Consider using pressure-regulating valves or other pressure compensation devices. Pressure compensation is crucial, particularly in systems with significant elevation changes. Remember, maintaining consistent pressure across all emitters, regardless of their position in the network, is essential for delivering uniform water distribution.

Proper installation of emitters is equally crucial. Incorrect placement can hinder water flow and create localized dry spots. Consistent depth and alignment of emitters should be the priority. This can require precision during installation. Ensure emitters are placed at the appropriate depth and aligned properly to provide the best possible water distribution. If emitters are misaligned, the water delivery will be inconsistent, possibly resulting in sections without the required hydration. The consequences of this include underperforming areas and a decrease in your overall yield.

Finally, regular monitoring and adjustments are vital for maintaining uniform water distribution. Observe your plants closely for signs of stress or overwatering, and adjust your irrigation schedule or emitter settings accordingly. Regular inspections will allow for the proactive detection and remedy of any issues that could negatively affect water distribution, such as clogged emitters or leaks in the system. This would also allow you to react promptly to any changing conditions. Implementing a regular schedule of inspections will help to ensure uniform irrigation and thus increase crop yields significantly. The long-term benefits of a well-maintained system easily outweigh the cost of regular inspections and potential adjustments.

Consistent monitoring and proactive problem-solving are crucial for optimizing water usage and ensuring healthy crop growth. Therefore, make careful, informed decisions throughout the design and implementation processes to establish a drip irrigation system that will serve your specific needs for many years to come. By attentively following the detailed steps outlined in this chapter, one can build a drip irrigation system that delivers exactly the right amount of water to each plant, optimizing water utilization and resulting in substantial improvement of crop yields.

3.3 Emitter Installation

Careful emitter installation is paramount for a successful drip irrigation system. The process begins with selecting the appropriate emitters based on factors already discussed: crop type, soil conditions, and water pressure. Remember, consistent water delivery to each plant is the ultimate goal. Mismatched emitters lead to uneven growth, wasted water and compromised yields. Avoid this by meticulously planning your emitter placement.

Next, consider the precise placement of each emitter. Precise positioning is crucial for optimal water distribution. For example, in row crops, emitters should be spaced evenly along the lines, ideally near the base of the plant but avoiding direct contact to prevent damage. For trees or shrubs, multiple emitters may be needed, distributed strategically around the drip line, the area where the roots are most concentrated. This targeted application minimizes runoff and evaporation losses, maximizing the water's effectiveness.

Accurate emitter placement requires careful planning and precise execution. Before installation, lay out the tubing and mark the locations for each emitter using a measuring tape and permanent marker. This precision minimizes errors and ensures uniform spacing. Using pre-punched tubing can expedite this process and provides consistent emitter placement throughout the system. Always double-check your measurements and emitter positioning to prevent later issues. This methodical approach ensures uniform water delivery to every plant.

Once the locations are marked, installing the emitters is relatively straightforward. If using in-line emitters, simply push them securely into the pre-punched holes in the tubing. Ensure they're inserted firmly and flush with the tubing surface to prevent leaks or uneven water flow. For pressure-compensating emitters, be extra careful to follow the manufacturer's instructions regarding installation. These emitters may require special tools for proper insertion and ensure consistent flow across varying pressures, maintaining even water distribution throughout the system.

After the emitters are installed, thoroughly inspect the entire system for leaks. A small leak can become a major issue over time. Carefully check each emitter connection and section of tubing. Use a low pressure test run to ensure all emitters are functioning correctly and water is evenly dispersed. If you encounter leaks, promptly repair them using appropriate sealing techniques, which could involve replacing faulty emitters, tightening connections, or patching holes in the tubing with designated repair kits. This step prevents water loss and ensures the system's long-term efficiency.

Finally, and this is extremely important for long-term system success, regularly check and maintain the emitters and the whole system. Clogged emitters are a common problem. This can be a simple solution if found early. Regularly flush the system with clean water to prevent clogging. It's crucial to make this a routine procedure. Addressing issues promptly minimizes disruptions and maintains the overall efficiency and productivity of the drip irrigation system. Consider using filters to prevent debris from entering the system. You've invested significant time and resources in creating this system; proper maintenance ensures a solid return. Your consistent attention to detail in emitter installation and upkeep ensures the success of your drip irrigation project, directly impacting your yields and efficiency. Careful planning and consistent maintenance directly translate to higher agricultural yields.

3.4 Pressure Regulation

Maintaining consistent pressure throughout your drip irrigation system is paramount for uniform water distribution. Uneven pressure leads to inconsistent watering, impacting crop growth and potentially wasting valuable resources. Imagine one section of your field receiving a deluge while another remains parched; this scenario highlights the critical need for precise pressure regulation.

Pressure regulators are your allies in this endeavor. These devices, strategically placed within your system, reduce the incoming water pressure to a level suitable for your emitters. Selecting the right regulator involves understanding your system's pressure and the requirements of your chosen emitters. Overlooking this step can lead to emitter damage or inefficient watering. Manufacturers usually provide specifications detailing the operating pressure range of their regulators, a detail you shouldn't ignore. Matching these specifications to your system's characteristics is crucial.

Pressure varies throughout the irrigation network due to elevation changes and pipe friction. The higher the elevation, the lower the pressure. Conversely, longer pipelines create more friction, reducing pressure downstream. The longer your pipeline, the more significant this effect becomes, impacting the uniformity of your irrigation. This uneven pressure distribution can easily result in under-watering at the far end of your system. To counter this, consider installing pressure compensating emitters. These clever devices deliver a constant flow rate regardless of pressure variations, ensuring consistent watering across your entire field.

Regular pressure monitoring is a key element of effective irrigation management. Using a pressure gauge, check the pressure at various points in your system. This simple step will reveal any pressure inconsistencies. Armed with this information, you can strategically adjust regulators or identify potential issues such as leaks or blockages. Regular checks prevent minor issues escalating into significant problems. This also helps you develop a better understanding of your system's pressure profile.

Beyond pressure regulators, proper system design plays a vital role in pressure management. The correct sizing of pipes is crucial. Using pipes that are too small restricts water flow, increasing pressure and potentially damaging emitters. Conversely, pipes that are too large might lead to unnecessary costs and increased pressure fluctuations. The selection of the correct diameter pipe is not only cost-effective but significantly impacts the system's efficiency. This crucial decision should always be based on accurate calculations factoring in pipeline length, desired flow rate, and elevation changes.

Proper installation also contributes to maintaining pressure. Ensure all connections are tight and secure to prevent leaks, which can significantly reduce pressure. Avoid sharp bends and kinks in your tubing, as these create pressure points that restrict water flow. A well-planned and expertly installed system will operate smoothly, minimizing pressure fluctuations. Even small leaks can significantly decrease the overall pressure in a long pipeline, thereby reducing efficiency.

Consider the location of your pressure regulators carefully. Place them strategically upstream of sections with significant elevation changes or long pipe runs. This ensures effective pressure reduction in areas prone to pressure drop. Incorrect placement of regulators can result in pressure imbalances, hindering the effectiveness of your irrigation system. For optimum performance, think ahead, plan and place them at strategic locations to counteract expected pressure variations along your network.

Finally, don't underestimate the value of a well-maintained system. Regular cleaning of filters and emitters will prevent blockages, helping maintain consistent pressure. Schedule routine inspections to identify any problems early. Proactive maintenance greatly increases the longevity of your system and minimizes the need for costly repairs. This is a small investment compared to the potential losses associated with irrigation failures resulting from neglected maintenance. A system that's well cared for runs efficiently and minimizes water waste.

3.5 Tubing & Connectors

Choosing the right tubing and connectors is crucial for a long-lasting, efficient drip irrigation system. The tubing's material dictates its lifespan and resistance to clogging; polyethylene (PE) is a popular choice for its flexibility, durability, and resistance to UV degradation. However, consider the specific needs of your crops and soil conditions when selecting the appropriate diameter and wall thickness. Thinner tubing is easier to handle, but thicker tubing is more resistant to pressure fluctuations. For longer runs, larger diameter tubing is often necessary to prevent excessive pressure drop, ensuring consistent water delivery throughout the system. Careful consideration of these factors ensures the success of your investment.

The selection of connectors is equally vital for system integrity. Connectors need to be compatible with your chosen tubing material and size; this ensures a tight, leak-free seal. Different connector types suit various applications: barbed fittings offer a secure mechanical connection, requiring no special tools, perfect for quick repairs and simple setups. Compression fittings, using nuts and ferrules, provide a stronger, more reliable seal, particularly suited for high-pressure systems or situations needing frequent disconnections. Thread fittings are useful when connecting to valves and other components with threaded ports, but ensuring a tight seal is crucial to avoid leaks. Therefore, always invest in high-quality connectors which suit the specific needs of your system's setup.

Planning the layout of your tubing network and the placement of your connectors requires foresight. Optimizing the system's design will minimize pressure drops and ensure even water distribution throughout the fields. Avoid sharp bends and kinks, which can restrict water flow, causing inefficient water distribution. Use connectors strategically to facilitate expansion or modifications to the system in the future. This planning stage is often overlooked, but it's crucial for optimizing irrigation efficiency. Well-planned tubing runs will help avoid problems further down the line, ensuring you maximize water usage.

Regular inspection and timely maintenance are essential. Check connectors regularly for leaks; a small leak can lead to significant water waste over time. Inspect tubing for damage or clogging. Regular flushing can remove sediments and debris that might accumulate and clog emitters or restrict water flow. This simple maintenance task significantly extends the life of your system and contributes to your long-term savings. Proactive maintenance is a far better investment than reactive repairs; a well-maintained system is more efficient, reliable, and cost-effective over the long run. This small investment in time ensures a worthwhile return.

Remember the importance of proper installation. Avoid over-tightening connectors, which might damage them; a properly installed connector does not require excessive force. When burying tubing, ensure proper depth and protection to prevent damage from farm equipment or weather conditions. Proper installation of all components is paramount for the lifespan of the system as well as the effectiveness of your water distribution. Thorough installation procedures lead to fewer maintenance issues.

Proper pressure regulation is essential for the longevity of your system and for consistent water delivery to your plants. Pressure regulators and filters prevent excessive pressure that can damage the emitters and tubing. Pressure sensors can help to monitor and adjust system pressure in real-time and are an important component for larger, more sophisticated drip irrigation systems. Without these measures, you may end up needing to replace components more frequently, potentially costing you more in the long run. Pressure regulation is not simply a luxury, it's an integral part of a well-designed and functional drip irrigation system.

Accurate calculations of water pressure and flow rate are crucial for selecting appropriate tubing and connectors. Use online calculators or consult with a professional to determine the correct sizes and materials to ensure consistent water delivery throughout your field. These calculations are more complicated than many novices assume, taking into account many variables, including pipe diameter, length, elevation changes and the chosen flow rate of your emitters. Incorrect calculations might lead to insufficient water delivery or conversely, to water damage to your crops.

The type of soil influences the choice of tubing. Sandy soil may require tubing that is more resistant to abrasion from shifting soil particles. Clay soils might require tubing with larger diameters to overcome the increased resistance to water flow. Understanding these soil specific attributes is crucial to ensure a more effective distribution of water to your plants. This is one reason why site assessments in the previous chapter are so critical to making good decisions throughout the drip irrigation process.

Lastly, consider the long-term cost-effectiveness of your choices. While cheaper options may seem appealing initially, they may require more frequent replacements and maintenance, ultimately costing you more in the long run. Investing in high-quality tubing and connectors will lead to a more durable, reliable, and ultimately more cost-effective system over the lifespan of your agricultural enterprise. Although the initial expense might be larger, the long term cost savings will be demonstrably evident.

4 Irrigation Scheduling & Management

Effective irrigation scheduling is crucial for maximizing the benefits of drip irrigation. It's not simply about applying water; it's about applying the right amount of water at the right time. Overwatering leads to wasted resources and potential root rot, while underwatering stunts growth and reduces yields. Understanding your specific needs is key. This requires a multifaceted approach incorporating several factors.

Soil moisture monitoring is paramount. Various methods exist, from simple feel tests to sophisticated sensors. A simple feel test involves digging a few inches into the soil to assess its moisture content. While this offers a basic understanding, technological advancements offer more precise measurements. Tensiometers directly measure soil water tension, providing a quantitative measure of soil dryness. These devices are more accurate than feel tests, leading to better irrigation decisions. Moreover, soil moisture sensors using various technologies, including capacitance and time-domain reflectometry, provide real-time data, which can be integrated into automated irrigation systems. Choosing the appropriate method depends on budget constraints, technological expertise, and the scale of your operation. Regardless of the method, regular monitoring is crucial for accurate scheduling.

Irrigation frequency depends heavily on several interconnected factors. Firstly, the climate plays a significant role. Hot, sunny days with high winds lead to increased evapotranspiration, requiring more frequent irrigation. Conversely, cooler, cloudy days require less frequent watering. Secondly, the type of soil dramatically impacts irrigation needs. Sandy soils drain quickly, requiring more frequent, albeit smaller, applications of water. Clay soils, on the other hand, retain water longer, necessitating less frequent irrigation. Thirdly, the growth stage of your crop influences water requirements. Young seedlings have lower water demands compared to mature plants at peak growth. It's important to tailor irrigation schedules to meet the specific needs of your crops at various developmental phases. Understanding these factors allows you to design a schedule that consistently provides optimal moisture levels.

Precise scheduling requires careful consideration of several interacting variables. This includes accurate measurement of rainfall amounts. Supplementing irrigation based on rainfall data ensures that you avoid unnecessary water application. Careful monitoring of weather patterns through local forecasts helps anticipate periods of high evapotranspiration and adjust the irrigation schedule accordingly. Using the combination of readily available data (weather forecasts, rainfall amounts), combined with regular soil moisture monitoring and an understanding of your soil type and crop needs allows you to create a precise schedule for optimal irrigation.

Developing a practical irrigation schedule should be an iterative process. Begin with an estimated schedule based on your climate, soil type, and crop. Then, closely monitor your plants and soil moisture levels to fine-tune the schedule. Observe plant growth, leaf color, and overall health for signs of either overwatering or underwatering. Adjust the frequency and duration of irrigation according to your observations. Remember that the optimal irrigation schedule is not static; it needs adjustment as conditions change throughout the growing season. Regular monitoring and adjustments ensure that your plants receive the precise amount of water they need for optimal growth.

Efficient irrigation management goes beyond mere scheduling. It involves preventative measures to avoid common problems, ensuring the longevity and efficiency of your system. Regular inspection and cleaning of emitters is crucial to avoid clogging, which can lead to uneven water distribution. This can often be achieved through flushing the system periodically, with the frequency depending on the quality of your water source. Regular checks for leaks help conserve water and prevent damage to the system. A proactive approach to maintenance and timely repairs ensures peak operational efficiency of the drip irrigation system, maximizing its lifespan and minimizing water waste.

These steps are crucial for optimizing the efficiency and effectiveness of your drip irrigation investment, leading to higher yields and a better return on investment.

4.1 Soil Moisture Monitoring

Precise soil moisture monitoring is paramount for efficient drip irrigation. Understanding your soil's water content allows you to schedule irrigation precisely, avoiding both underwatering and overwatering, which can harm yields and waste precious resources. This isn't just about guesswork; it's about data-driven decision-making.

Several methods exist for monitoring soil moisture. Simple, time-tested techniques include feeling the soil's texture - dry soil feels crumbly, while moist soil is pliable. However, this method is subjective and lacks precision. More accurate methods involve using a soil moisture meter, a handy tool that directly measures the volumetric water content of the soil. Different meters use various technologies, such as time-domain reflectometry (TDR) or capacitance sensors. Choose a meter appropriate for your needs and soil type, considering factors like depth and ease of use.

Beyond handheld meters, advanced technologies offer comprehensive solutions. Tensiometers measure the soil's water tension, giving you insights into the soil's water potential and the force with which water is held by the soil particles. These provide a more nuanced understanding of the soil's water availability than simple volumetric content. Then there are soil moisture sensors that can be integrated into a larger irrigation system for automated, real-time monitoring and control. These often wirelessly transmit data to a central hub, allowing for remote monitoring and adjustments to irrigation schedules.

Regardless of the method you choose, proper calibration and regular maintenance are essential. Handheld meters may need calibration against known moisture levels to ensure accuracy. Sensors may need periodic cleaning to prevent inaccuracies caused by debris or mineral buildup. Consistent maintenance is key to obtaining reliable and precise data. Data logging, whether manually recording readings or relying on automated sensor systems, is crucial to building a history of your soil's water content.

This historical data becomes invaluable when you analyze trends and patterns. You'll learn how your soil retains water under different conditions and how various weather patterns affect soil moisture. This allows you to fine-tune your irrigation schedule, adjusting for variations in rainfall and evapotranspiration rates. Remember, you're not simply reacting to current conditions; you're proactively managing your irrigation strategy based on long-term observations. This proactive approach is the key to achieving optimal water usage and maximizing crop yields.

Consider the specific needs of your crops when interpreting soil moisture data. Different crops have different water requirements at various growth stages. For instance, young seedlings typically need more frequent watering than mature plants. Understanding these variations is crucial for accurate irrigation scheduling, ensuring that plants receive the appropriate amount of water throughout their life cycle. You may need to adjust your monitoring frequency based on the crop's growth stage and the soil's characteristics. Moreover, soil type significantly influences moisture retention. Sandy soils drain quickly, requiring more frequent but lighter irrigation compared to clay soils, which hold water for longer periods.

Effective soil moisture monitoring also enables better integration with other elements of your drip irrigation system. This data can inform the selection and placement of emitters, considering the water-holding capacity of your soil type and the water requirements of your crops. The data helps optimize your fertilization schedule and techniques - nutrient uptake is directly related to adequate soil moisture. By carefully monitoring the moisture levels, you can ensure nutrients are properly absorbed by the plants, avoiding nutrient runoff or nutrient deficiencies that can hinder growth. Furthermore, integrating soil moisture monitoring with weather forecasts allows for predictive irrigation scheduling, enabling you to proactively adjust your irrigation strategy in anticipation of rainfall or extreme temperatures.

In conclusion, embracing precise soil moisture monitoring transforms your approach to drip irrigation from a reactive strategy to a proactive, data-driven management system. The investment in monitoring tools and techniques pays off through water conservation, enhanced crop yields, and a deeper understanding of your soil and its interaction with your crops. Remember, consistent monitoring and analysis are key to mastering this aspect of drip irrigation and reaping its significant rewards. Don't underestimate the value of accurate, ongoing data; it's your guide to maximizing your farm's potential.

4.2 Irrigation Frequency

Determining the ideal irrigation frequency for your drip system is crucial for maximizing crop yields while conserving water. It's not a one-size-fits-all solution; rather, it's a nuanced dance between several factors, requiring careful observation and adaptation. Ignoring this delicate balance can lead to water stress, reduced yields, or wasteful overwatering. Success hinges on understanding your specific conditions.

Soil type significantly impacts irrigation frequency. Sandy soils, with their excellent drainage, require more frequent, albeit smaller, irrigations to prevent rapid water depletion. Clay soils, on the other hand, retain moisture longer, demanding less frequent but potentially larger irrigation events. Knowing your soil's water holding capacity, often determined through soil tests or experience, is paramount for accurate scheduling. This information allows you to tailor irrigation to match your soil's specific needs. The frequency of your watering will reflect the nature of your soil and the crop you intend to grow.

Crop water requirements are another critical aspect. Different plants have different thirst levels. A water-intensive crop like tomatoes will demand more frequent watering than a drought-tolerant species. Understanding the evapotranspiration rate—the combined process of evaporation from the soil surface and transpiration from the plants—for your chosen crop provides a valuable indicator of water needs. Consult agricultural extension services or research-based resources for crop-specific ET data. This, coupled with your understanding of soil composition, allows you to create a customized schedule that promotes the optimal growth of the intended plants. You will avoid both under and over-watering.

Weather patterns influence irrigation scheduling dramatically. Hot, sunny days accelerate evaporation and transpiration, requiring more frequent irrigation. Conversely, cooler, cloudy days may allow you to extend the time between waterings. Regularly monitoring weather forecasts, coupled with on-site observations of soil moisture, helps in making informed irrigation decisions. Consider using weather stations or rainfall gauges to accurately assess precipitation and environmental conditions. This will enhance your capacity to predict and tailor your irrigation plan to meet the needs of your crops.

Monitoring soil moisture levels directly is a pivotal technique. Soil moisture sensors, available in various forms, provide real-time data on soil water content. This removes guesswork, enabling more precise irrigation scheduling. Alternatively, you can use the 'feel' method: dig a few inches into the soil to check its moisture content or use a simple moisture meter. These simple steps allow for a more precise understanding of how much water is present and, consequently, the best time to water.

Technological advancements offer sophisticated irrigation management tools. Smart irrigation controllers use weather data, soil moisture sensors, and even plant health indicators to automate irrigation schedules. While these systems are not strictly necessary for smaller operations, they can significantly enhance efficiency and water conservation on a larger scale. While the cost of implementation may be a concern, the increased precision and automated approach can outweigh the expenses over time and save you a considerable amount of water.

Finally, remember that your irrigation schedule should be dynamic, not static. Regularly observe your plants for signs of water stress (wilting, leaf curling) or overwatering (yellowing leaves, stunted growth). Adjust your irrigation frequency accordingly. This adaptive approach ensures your plants receive the optimal amount of water throughout their growing cycle. Regular monitoring of your plants' health is essential for adapting your irrigation schedule. Constant observation will eventually allow you to predict patterns and adjust accordingly. This practice may save you significant amounts of water and money.

5 Water Quality & Fertigation

Maintaining optimal water quality is paramount for successful drip irrigation and efficient fertigation. High salinity levels, for example, can severely damage plant roots, reducing yields and potentially causing irreparable harm to your system. Regular water testing, identifying the specific ions causing issues (sodium, chloride, bicarbonate are common culprits), and implementing appropriate mitigation strategies - such as using a reverse osmosis system or blending water sources - are essential steps. Understanding your water's chemical composition is the first step towards a healthy system and thriving crops.

Conversely, low water quality can manifest in different ways. For instance, excessive sediment can lead to blocked emitters, compromising uniform water distribution and potentially stressing plants in sections of your field. Regular filtration, with appropriately sized filters for your system's needs and your water's unique characteristics, prevents this. Consider using multi-stage filtration systems for particularly problematic water sources, ensuring that your precious water is delivered to your crops consistently and effectively.

Fertigation, the simultaneous application of water and fertilizers through the drip irrigation system, offers a precision approach to nutrient management. However, it demands careful planning and execution. The precise concentration and types of fertilizers depend greatly on your crop's specific needs throughout its growing cycle. Excessive fertilizer application can lead to salt buildup in the soil, a damaging condition that can easily be avoided with thoughtful planning. Consulting with agricultural experts and utilizing soil tests to determine the exact nutrient requirements of your crops is advisable, ensuring optimal plant health.

Accurate fertilizer injection is also critical. Proportioners, which accurately mix fertilizers into the irrigation water, should be chosen according to your specific irrigation needs and flow rates. Regular calibration of these devices is essential to maintain consistent fertilizer application. Improper calibration can result in uneven nutrient distribution or excessive fertilizer application, negatively affecting the growth and health of your crops, and potentially damaging the irrigation system itself. It is vital to regularly check and maintain these components, acting swiftly to address any inaccuracies.

The selection of appropriate fertilizers is just as important as their application. Water-soluble fertilizers are generally preferred for drip irrigation due to their ease of dissolution and compatibility with the system. However, even water-soluble fertilizers can lead to clogging if not carefully managed. Consider the solubility and particle size of your chosen fertilizers, understanding the potential challenges each might pose to your system. Selecting the right fertilizer, in the right amounts, and applying it in the right way is an art form, an art form essential to maximize your yields and your system's life.

Furthermore, regular monitoring of the irrigation system is crucial. Observe emitters closely for signs of clogging. Check the pressure in the system regularly to identify potential issues before they escalate. Early detection of problems and prompt remediation are key to minimizing system downtime and ensuring a healthy environment for your crops. It is far better to act preventatively than reactively, saving you time, money and crop yields. Regular inspections will increase your efficiency and allow for early intervention before small issues become major problems.

Beyond routine maintenance, consider incorporating advanced monitoring techniques. Soil sensors can provide valuable data on soil moisture levels and nutrient concentrations, allowing for more precise irrigation and fertilization scheduling. Data logging systems can track operational parameters, helping you to fine-tune your system's performance over time. The more you understand your system's behavior, the better equipped you are to optimize its efficiency and effectiveness, ultimately improving yields and reducing water waste. Investing in these technologies offers significant returns in terms of precision and overall system management.

Finally, remember that fertigation is a dynamic process. Adapt your nutrient management strategies based on plant growth stages, soil test results, and environmental conditions. Regular monitoring and adjustment are crucial to ensuring consistent and appropriate nutrient supply. Flexibility and responsiveness to changing conditions are key characteristics of successful fertigation. This requires vigilance, observation and a willingness to adapt your techniques. Your crops will reflect the care and attention you give to optimizing their nutrition and hydration.

5.1 Nutrient Management

Efficient nutrient management is paramount in drip irrigation, maximizing crop yields while minimizing environmental impact. Understanding the specific nutrient requirements of your chosen crop is the foundational step. Conduct a thorough soil test to determine existing nutrient levels, identifying deficiencies or excesses that may influence your fertilization strategy. This initial assessment allows you to tailor your nutrient plan to the crop's needs, preventing over-fertilization, which can harm the environment through nutrient runoff and negatively impact crop health. A balanced approach, focusing on the right nutrients in the right amounts at the right time, is key to success. This is especially important with drip irrigation because of the targeted nature of nutrient delivery.

The choice of fertilizer is equally crucial. Water-soluble fertilizers are best suited for drip irrigation, dissolving readily and ensuring even distribution through the system. Avoid fertilizers that could clog emitters; those with high salt concentrations or those containing large insoluble particles should be avoided. Opting for slow-release fertilizers might seem attractive for convenience, but these are usually unsuitable for drip systems due to their potential to create blockages within the tubing and emitters. Careful consideration of the fertilizer's composition is vital, ensuring it aligns with the specific needs identified through your soil analysis. The correct balance of macronutrients (nitrogen, phosphorus, and potassium) and micronutrients is essential for optimal plant growth and overall yield.

Precise application techniques are vital to avoid nutrient loss and ensure effective uptake by the plants. Fertigation, the simultaneous application of water and fertilizer, is a highly efficient method within drip irrigation. The precise control afforded by drip systems allows for careful regulation of fertilizer concentrations, minimizing waste and optimizing nutrient delivery to the roots. Start with lower concentrations than recommended on the fertilizer packaging and monitor plant response; this phased approach helps prevent burning the plants or causing salt buildup in the soil. Regular monitoring of the nutrient levels in your irrigation water is crucial, especially after any adjustments to the fertilization regime. This prevents imbalances, which might lead to nutrient deficiencies or toxicities in plants, thus impacting yields and overall crop health.

Regular monitoring of the plants themselves is vital for understanding how they're responding to the applied nutrients. Visual cues, such as leaf color and overall growth, are good initial indicators of nutritional status, but more detailed techniques, such as tissue analysis, are also useful. Tissue analysis provides a quantitative assessment of nutrient content in the plant itself, providing precise data for fine-tuning your fertilization strategy. This advanced technique can highlight subtle deficiencies or toxicities that might otherwise be missed, offering invaluable insights for informed decision-making. By combining visual observations with tissue analysis, you can gain a comprehensive understanding of your plant's nutritional status and adapt your fertilization strategy accordingly.

Remember that water quality plays a significant role in nutrient delivery. High levels of salts or certain minerals in the irrigation water can interact with the fertilizer, impacting its effectiveness or causing further complications within the system. Regular water quality testing is as important as monitoring the soil and plant nutrient levels. Water quality problems can lead to emitter clogging, reduced nutrient uptake, and ultimately, reduced crop yields. Taking preventative measures, such as using appropriate filters, will reduce clogging and ensure long-term efficiency and effectiveness of the irrigation system. Addressing water quality issues early on, through adequate filtration and regular monitoring, is crucial for maintaining the longevity and efficiency of your drip irrigation system, thus maximizing both yield and efficiency.

Consistent monitoring, coupled with careful adjustments based on plant response and soil conditions, is a constant process. It's not a "set it and forget it" situation. The nutrient requirements of a crop constantly change throughout its growth cycle. Therefore, regular monitoring and adjustments are vital. This adaptive approach to nutrient management, which includes regular tissue and soil testing, plus ongoing assessment of water quality and plant response, will lead to the highest yields and optimized nutrient use in your drip irrigation system. Effective nutrient management is a key element in achieving maximum yields and enhancing overall sustainability. This cyclical process of monitoring, adjusting, and refining your strategy ensures you achieve optimal performance and minimize environmental consequences.

5.2 Fertilizer Application Techniques

The successful implementation of fertigation, the application of fertilizers through a drip irrigation system, hinges on careful planning and execution. Understanding the characteristics of your chosen fertilizer is paramount. Is it soluble? Does it contain components that might clog emitters? Thorough research into fertilizer compatibility with your drip system is crucial, preventing costly blockages and equipment damage. Remember, prevention is always cheaper than cure. Moreover, consider the specific nutrient needs of your crop at various growth stages. A detailed understanding of these requirements allows for precise tailoring of the fertilizer solution, maximizing yield while minimizing environmental impact. Improper fertilization can lead to nutrient deficiencies or excesses, both detrimental to plant health and overall productivity. Begin by preparing the fertilizer solution according to the manufacturer's instructions. Accurate measurement is essential here; using calibrated equipment ensures precise nutrient delivery. Always dissolve the fertilizer completely in water before introducing it into the irrigation system, to avoid clogging the emitters. This seemingly simple step is often overlooked, leading to significant issues downstream. Introduce the solution slowly, monitoring the pressure within the system to prevent damage. Consistent monitoring is key throughout the entire process, checking for any anomalies or signs of clogging. Remember to maintain detailed records of fertilizer application, including dates, concentrations, and quantities. This provides valuable data for future planning and fine-tuning of your fertigation strategy, allowing for continuous improvement in your irrigation practices.

The placement of fertilizer injectors within the drip irrigation system is a critical design aspect. Their location should be chosen strategically to ensure even distribution of nutrients across the entire field. Factors to consider include the system's pressure, flow rates, and the proximity of the injectors to the emitters. Incorrect placement could lead to uneven nutrient distribution, harming crop growth and yield. Regular monitoring of the fertilizer solution's concentration and flow is essential, ensuring consistent delivery. If you detect fluctuations, immediately investigate potential sources of blockage, such as sediment build-up or fertilizer precipitates. Proactive monitoring can prevent larger, more expensive, and time-consuming problems later on.

Beyond the initial setup, the ongoing maintenance of your fertigation system is vital for sustained success. Regular flushing of the lines is crucial to prevent the build-up of fertilizer residues. These residues, if left unchecked, can crystallize and severely impact emitter functionality, reducing their lifespan and compromising irrigation uniformity. Remember to clean the injectors and filters according to a scheduled routine, to maintain optimal performance. Furthermore, monitor the health of your crops regularly. Note any unusual growth patterns, color changes, or signs of nutrient deficiencies or toxicity. These observations can indicate issues within your fertigation system, allowing for prompt adjustments to your fertilization strategy and prevention of wider problems. Consider conducting periodic soil tests to evaluate nutrient levels. This provides data-driven insights into the effectiveness of your fertigation program, enabling you to make informed decisions for optimization. By integrating these techniques, your drip irrigation system will become a tool for precision agriculture, maximizing efficiency, yield, and profitability while minimizing environmental impact.

In conclusion, proper fertilizer application techniques within a drip irrigation system are integral to maximizing crop yield and profitability. Careful planning, accurate measurement, strategic injector placement, and regular maintenance are crucial for ensuring the success of your fertigation program. Remember that meticulous record-keeping, along with regular monitoring and soil testing, are essential elements of this process. By following these guidelines, you can harness the power of fertigation to optimize your irrigation efforts and achieve significant improvements in crop production. This holistic approach translates to efficient water and fertilizer usage, reducing waste and maximizing returns on your investment. Effective fertigation demonstrates a dedication to sustainable and efficient farming practices, reflecting a commitment to both environmental protection and economic success. Through careful planning and diligent monitoring, you'll significantly enhance the efficiency and effectiveness of your drip irrigation system, guaranteeing healthier crops and ultimately, a more profitable agricultural operation.

5.3 Water Quality Issues

High salinity significantly impacts drip irrigation systems. Excessive salts in the irrigation water can accumulate in the soil, hindering water uptake by plant roots and ultimately reducing crop yields. This salt buildup, a phenomenon known as soil salinity, is particularly detrimental to sensitive crops. The resultant osmotic stress prevents efficient water absorption, even if sufficient water is supplied. Moreover, certain salts can clog emitters, reducing the efficiency of the entire system and requiring more frequent maintenance. Careful monitoring of water salinity is crucial, utilizing appropriate testing methods to gauge its concentration and its potential effects on the crops and the irrigation equipment. Regular flushing of the system can help mitigate this problem, but it's more effective to address the problem at its source, ensuring that the water used for irrigation is suitably low in salinity.

Addressing high pH levels in irrigation water requires a proactive approach. A high pH can reduce nutrient availability to plants, impacting their overall health and productivity. This is because the high pH may cause certain essential nutrients, like iron and zinc, to become less soluble and less accessible for plant uptake. Furthermore, high pH can increase the likelihood of calcium and magnesium carbonate precipitation, potentially leading to emitter clogging. Soil testing to determine current pH levels is a critical first step, allowing for informed decisions about adjustments to irrigation practices. Acidification of irrigation water using readily available products, like phosphoric acid, can effectively neutralize the alkalinity, making essential nutrients available to the plant. Monitoring pH levels and plant health after implementing these measures is important to confirm their effectiveness. It's essential to avoid drastic changes, adjusting the pH gradually to avoid negative impacts on soil chemistry and plant life.

Suspended solids, like silt, clay, and organic matter, are another major challenge in drip irrigation. These particles can easily clog emitters, reducing water flow and leading to uneven irrigation. The problem is exacerbated in water sources prone to high sedimentation, such as surface water bodies after periods of heavy rainfall. Effective pre-filtration is critical in mitigating this issue. Various filtration techniques, such as sand filtration and screen filtration, can effectively remove suspended solids, ensuring the efficient functioning of the drip irrigation system. The choice of filtration method should be guided by the characteristics of the water source and the required level of filtration. Regular maintenance, including backwashing and cleaning of filters, is crucial for maintaining their efficacy and preventing clogging. Ignoring this aspect can negate the benefits of drip irrigation due to the cost and labor involved in system repairs and replacements.

Biological contamination, including bacteria, fungi, and algae, can cause significant problems in drip irrigation systems. These microorganisms can proliferate within the tubing and emitters, leading to clogging, reduced flow rates, and potential damage to the system's infrastructure. Poor water quality from contaminated sources or stagnant water in the lines are potential vectors. The use of water disinfectants, such as chlorine or hydrogen peroxide, can effectively control biological contamination. However, it's vital to be aware of the potential impact of these disinfectants on plant health and to use them only as directed. Regular flushing and maintenance can also help minimize the risk of biological growth. It's always crucial to follow local guidelines and regulations regarding safe and responsible use of such chemicals. Prevention is key, choosing water sources known for their purity can save the time and effort necessary to resolve more serious issues.

Iron and manganese are prevalent in many water sources, but their presence can significantly affect drip irrigation systems. These metals, when dissolved in water, can precipitate as insoluble oxides, particularly when exposed to oxygen or changes in pH. This precipitation can lead to emitter clogging, particularly in areas with low flow rates. Water testing is essential for determining iron and manganese concentrations, helping to decide on the most effective remediation strategy. Specific water treatment techniques like oxidation and filtration are available to remove or reduce these metal levels in water used for irrigation. Selecting appropriate materials for the system that are resistant to corrosion caused by high iron and manganese is also crucial. This long-term consideration can reduce the frequency and cost of system maintenance and replacements. Regular monitoring of iron and manganese levels ensures that corrective measures can be taken proactively to maintain efficient irrigation.

6 Troubleshooting & Maintenance

Regular maintenance is paramount for the longevity and efficiency of your drip irrigation system. Think of it as preventative medicine for your crops; addressing small issues before they become major headaches. A proactive approach will save you time, money, and frustration in the long run. Start by visually inspecting the entire system at least once a month, checking for any signs of leaks, damaged tubing, or clogged emitters. Look closely at connections, ensuring they're secure and free of leaks. Pay particular attention to areas prone to wear and tear, such as those near machinery or where the tubing is exposed to extreme temperature fluctuations. Remember, early detection is key.
Regular flushing is crucial for preventing emitter clogging. This involves running clean water through the system at a higher-than-normal pressure for a short period. This helps dislodge any sediment or debris that may have accumulated. The frequency of flushing depends on your water source's quality; hard water or water rich in sediment requires

more frequent flushing—perhaps every couple of months. For softer water sources, flushing every 6 months may suffice. Always check your system's manufacturer recommendations for specific flushing guidelines. Remember to properly drain the system after flushing to prevent water stagnation.

Clogged emitters are a common problem. They restrict water flow, leading to uneven watering and reduced yields. Several methods can effectively clear blockages. You can try using a small wire or a specialized emitter cleaning tool to clear the blockage. Alternatively, you can carefully dismantle the emitter, clean it with a soft brush and water, and then reassemble it. If the clogging is severe or widespread, consider replacing the affected emitters. Remember to always use the correct tools and follow safety precautions to avoid damage to the emitters. Don't force anything; if you're struggling, replacement is a far more efficient solution.

Identifying and repairing leaks promptly is essential. Small leaks can escalate into significant water loss and system damage if left unchecked. Inspect all connections and tubing regularly for signs of leakage, such as damp patches or water trickling. Minor leaks can often be fixed by tightening connections or patching small holes with specialized repair tape. However, for more significant leaks, replacing damaged sections of tubing or connectors might be necessary. Remember, maintaining proper pressure within the system is crucial for minimizing leaks. Regularly check and adjust pressure regulators as needed.

Beyond regular inspections and flushing, consider implementing a more comprehensive maintenance schedule. This may involve a thorough cleaning of the entire system annually. This process often involves dismantling portions of the system to thoroughly remove any debris or sediment buildup. A comprehensive cleaning is more involved and may require specialized cleaning tools or professional assistance. In regions with freezing temperatures, winterizing your system is vital to prevent damage from frost. This typically involves completely draining the system and storing sensitive components indoors until the spring.

Remember, investing in preventative maintenance saves you significant resources and ensures optimal irrigation system performance. Consistent attention to details and a proactive approach will guarantee healthier crops and a longer lifespan for your equipment. Proper documentation of maintenance activities, including dates and types of repairs, is highly beneficial for future reference and troubleshooting. Such records can assist in identifying patterns, predicting future problems, and optimizing maintenance schedules based on actual system performance and environmental conditions. A well-maintained drip irrigation system is an investment that pays off handsomely in increased yield, reduced water usage, and long-term cost savings. Don't underestimate the value of careful maintenance in the success of your agricultural operation. The effort you put in now yields far greater returns in the years to come.

6.1 Clogged Emitters

Clogged emitters are a common frustration in drip irrigation, significantly impacting water distribution and crop health. Identifying the cause is the first crucial step in resolving the issue. Several factors contribute to this problem; understanding them allows for preventative measures and efficient remediation. Poor water quality, with high sediment or mineral content, is a major culprit. These particles accumulate within the emitters, gradually restricting or completely blocking water flow. Regular filtration, properly sized for the system's capacity and water source characteristics, is paramount in preventing this.

The selection of appropriate emitters for the specific soil type and crop requirements also plays a significant role. Using emitters with smaller orifices in sandy soils, for example, can lead to increased clogging. Conversely, larger orifices in heavy clay soils may not provide sufficient pressure to overcome the resistance. Careful consideration during the design phase, ensuring compatibility between emitter size, soil type, and system pressure, minimizes clogging risks. Furthermore, the use of self-flushing emitters, which incorporate mechanisms to periodically expel accumulated debris, can enhance system longevity.

Another critical factor often overlooked is the installation process itself. Incorrect installation, leaving debris inside the tubing or emitters during placement, can lead to rapid clogging. Careful attention to detail during installation, including flushing the lines after installation, minimizes the risk of introducing contaminants that could obstruct the emitters. Thorough cleaning of the tubing and emitters before installation, combined with a system-wide flushing procedure using clean water, is a simple yet highly effective preventative measure.

Beyond these factors, biological growth, such as algae or bacteria, can also obstruct emitters, especially in warmer climates or systems with stagnant water. Regular maintenance, including flushing the system with a mild cleaning solution, helps mitigate this. The frequency of this cleaning depends on several factors, including climate, water quality, and the type of emitters used. A schedule should be established and followed consistently to minimize biological growth and ensure optimal emitter performance.

Finally, pressure fluctuations within the system can exacerbate clogging. Insufficient pressure may allow sediments to settle within emitters, while excessive pressure can force debris into the emitters more forcefully. Regular pressure checks and adjustments, using pressure regulators where needed, ensure consistent flow and reduce the risk of clogging. Understanding your system's pressure requirements and employing appropriate pressure regulation devices are essential for long-term efficiency and reduced maintenance. Remember, proactive maintenance, combined with careful system design and emitter selection, significantly extends the lifespan of your drip irrigation system and prevents costly downtime due to clogged emitters, allowing for the best use of water and ensuring optimal crop yields. A little preventative care goes a long way. Ignoring these issues can lead to substantial losses in yield, and therefore profit. Taking the time to ensure proper water flow will protect your investment.

6.2 System Leaks

Identifying a leak in your drip irrigation system can be frustrating, but a systematic approach is key. Start by visually inspecting the entire system, looking for obvious signs like wet patches on the ground, consistently damp soil in unexpected areas, or even visible breaks or cracks in the tubing. Don't rush this initial visual check; take your time. A careful eye can often pinpoint the source quickly. Pay close attention to connections, as these are often points of failure.

Next, utilize the power of water pressure. If you have a pressure gauge installed, monitor the pressure at various points along the main line. A significant drop in pressure between two points indicates a leak somewhere in between those points. Divide and conquer, checking sections systematically until you isolate the problem. If you don't have a gauge, consider investing in one; it's an invaluable tool for preventative maintenance and rapid leak detection. It will save you time and water in the long run. Remember to always turn off the system before performing any maintenance or repairs.

Once you've narrowed down the potential leak area, a more detailed inspection is needed. Carefully examine each section of tubing, connector, and emitter within that zone. Gently press along the tubing to feel for any unusual softness or bulging, indicating a potential puncture. Look for cracks, especially near bends and junctions where stress is most concentrated. For stubborn leaks, consider using a low-pressure water spray to help pinpoint the exact location; the escaping water will become visible under the stream. Remember safety first; always wear appropriate protective gear during these inspections.

Repairing a leak depends on its location and severity. Minor punctures in the tubing can often be repaired with specialized self-healing tape, designed specifically for this purpose. This tape conforms to the tubing, providing a watertight seal. Ensure the tubing is thoroughly clean and dry before applying the tape to maximize its effectiveness and longevity. For more significant damage or leaks at connections, you might need to replace the affected section of tubing or connector completely. This involves cutting out the damaged portion and carefully joining the new section with appropriate connectors, ensuring a secure, leak-free fit.

Larger leaks, especially those originating from major components like valves or filters, might require more extensive repairs. You might need to consult a professional irrigation specialist. These larger-scale repairs often necessitate specialized tools and expertise to ensure the system's integrity and correct functioning. Don't hesitate to reach out to a professional if you're unsure about the complexity of the problem.

Prevention is always better than cure. Regular system maintenance is vital for preventing leaks. A thorough inspection after each irrigation cycle is a good practice. Look for any signs of weakness or deterioration. This will give you an opportunity to address minor issues before they become major problems. The best way to avoid bigger problems is paying attention to the smallest issues. Schedule regular flushing of the system to clear any debris that could clog emitters and cause unnecessary back pressure, leading to potential leaks. Remember, proactive maintenance can save you significant time, money, and water in the long term.

Beyond regular inspections and flushing, consider the environmental factors. Extreme temperature fluctuations can cause stress on the tubing and connectors. Freezing temperatures can cause the water inside to expand and potentially crack the tubing. Similarly, prolonged exposure to direct sunlight can degrade the material over time, making it more susceptible to cracking or punctures. Protecting your system from harsh weather conditions, such as through shading or appropriate insulation in colder climates, can extend its lifespan and reduce the likelihood of leaks.

Finally, documenting your system is crucial. Keep a detailed record of the system's layout, component specifications, and maintenance history. This documentation will be invaluable during troubleshooting, especially for large or complex systems. A clear, comprehensive record allows you to identify potential problem areas more quickly and effectively, saving you time and resources when dealing with leaks or other issues. It also facilitates efficient repairs and helps you track system performance over time, leading to improved management and water use efficiency. Taking pictures as you lay out the system can be a great way to create such a record.

6.3 Regular Maintenance

Regular maintenance is paramount for the longevity and efficiency of your drip irrigation system. Neglecting this vital aspect can lead to decreased performance, crop stress, and ultimately, reduced yields. Think of it like servicing your car; regular checks prevent major breakdowns later on. This proactive approach saves you time, money, and frustration in the long run, ensuring your investment remains profitable.

A thorough visual inspection should be your first step. Walk the entire system, carefully examining all components. Look for any signs of physical damage, such as cracks in tubing, loose connections, or broken emitters. Pay close attention to areas prone to wear and tear, such as junctions and bends in the pipeline. These visual checks can quickly identify potential problems before they escalate. A damaged section, if left unaddressed, can lead to water loss, uneven distribution, and significant system failure requiring expensive repairs later on.

Cleaning is another crucial element of regular maintenance. Clogged emitters are a common issue, particularly in areas with high mineral content in the water source. Regular flushing of the entire system is recommended, preferably at the end of each growing season. This involves opening the mainline valve and allowing water to flow freely through the system for a sufficient period, dislodging sediment and debris. Consider using a specialized cleaning solution to tackle stubborn blockages, always following the manufacturer's instructions. This simple procedure can drastically extend the lifespan of your emitters, ensuring optimal water delivery to your plants.

Pressure regulation needs careful attention. Inconsistent water pressure can lead to uneven watering and potential damage to the system's delicate components. Regularly check pressure gauges at different points within the system to ensure consistent flow. If pressure fluctuations are detected, investigate the cause, addressing any leaks or blockages. Maintaining proper pressure is crucial for efficiency, extending the life of your system while ensuring even distribution of water and nutrients. A slight pressure discrepancy may seem unimportant, but it may add up to significant water and nutrient loss over time.

Beyond the physical system, consider the water quality. Regular water testing is advised to identify potential issues like excessive mineral content, which could lead to clogging or damage to components. This information allows you to adopt appropriate preventative measures, like more frequent cleaning or considering water filtration upgrades. This preventative approach is crucial for preventing long-term issues and maximizing the life of your irrigation setup. Early intervention is significantly cheaper than replacing entire sections of your system.

Documenting maintenance activities is highly recommended. Keep a detailed log of all inspections, cleaning procedures, repairs, and any issues encountered. This detailed record provides valuable insights into the system's performance, identifying patterns and trends that might indicate recurring problems or areas needing improvement. This information will prove invaluable in planning future maintenance and for troubleshooting issues in your system, helping you to avoid costly surprises and delays during critical growth periods.

Finally, remember that the specific maintenance schedule depends on various factors, such as climate conditions, water quality, and the type of crops being irrigated. Adjust your maintenance plan accordingly, prioritizing those tasks most relevant to your specific situation. Regularly consulting your system's manual, and seeking expert advice when needed, is highly recommended to optimize this crucial aspect of your drip irrigation strategy. A well-maintained system provides superior performance, increases yields, and ensures a high return on your initial investment.

7 Advanced Techniques & Applications

Smart irrigation systems represent a significant advancement in water management, moving beyond traditional scheduling. These systems leverage sensors to monitor soil moisture, weather conditions, and even plant health indicators in real-time. This data feeds into sophisticated algorithms that optimize irrigation schedules, adjusting water delivery based on actual needs rather than pre-programmed cycles. The result is a considerable reduction in water waste, improved crop yields, and minimized environmental impact. Imagine a field where every plant receives precisely the amount of water it requires, all managed autonomously, reacting dynamically to environmental fluctuations. The precision afforded by these systems allows for targeted watering based on the varying needs of individual plants within a field, which ensures optimal resource allocation and reduces the risk of overwatering or underwatering. The economic benefits can be substantial, as water costs are significantly reduced and labor requirements are minimized due to automation. Moreover, sophisticated systems enable remote monitoring and control, permitting adjustments and oversight from any location with internet access. This enhanced flexibility facilitates swift responses to unexpected weather patterns or system malfunctions, guaranteeing consistent irrigation performance. The integration of smart technologies is not merely an upgrade; it is a transformation in how we approach irrigation management.

Precision agriculture, enabled by advanced drip irrigation, promotes site-specific crop management. This contrasts with conventional methods that often treat entire fields uniformly, regardless of variations in soil conditions, topography, or plant health. Utilizing techniques like variable rate irrigation, we can deliver different amounts of water to specific areas based on their unique requirements. This requires detailed mapping of the field, integrating data from sensors and GIS systems to precisely determine the water needs of specific zones. This detailed approach optimizes water and fertilizer application, maximizing resource utilization and reducing environmental impact. We might identify zones with poor drainage or higher salinity, then adapt irrigation schedules or even apply specific amendments precisely where needed. This detailed management significantly minimizes input costs, reduces waste, and fosters sustainable agriculture practices. By addressing variability within a field, precision agriculture maximizes yields and minimizes resource consumption. It is a paradigm shift from generalized application to highly targeted and customized management.

Consider a vineyard in California struggling with drought. By implementing a smart irrigation system with soil moisture sensors and weather data integration, the vineyard manager could precisely adjust irrigation based on real-time conditions. This ensured the vines received adequate water during crucial growth stages, while preventing water stress and maximizing grape yield. The system automatically reduced irrigation during rainy periods, saving water and preventing soil saturation. Moreover, data collected by the system helped the manager optimize fertilization, applying nutrients only when and where needed. This approach improved not only water efficiency but also reduced fertilizer costs while safeguarding the environment. Similar success stories exist across various crops and climates, showcasing the adaptability and efficacy of advanced drip irrigation. These real-world examples demonstrate its potential to transform agricultural practices towards sustainability and increased profitability. Careful analysis of such case studies is crucial for adapting these strategies effectively to other settings. These scenarios highlight the powerful synergy between technology and sustainable agriculture.

The integration of advanced techniques extends beyond single farms. Larger-scale projects demonstrate the scalability and impact of these strategies. For example, a community irrigation project in a developing nation could use smart irrigation and precision agriculture to optimize water distribution across a collective of small farms. The shared network would allow efficient allocation of resources and promote equitable access to water, fostering community development and economic growth. This showcases the potential of advanced drip irrigation in addressing global challenges related to food security and sustainable water management. This collaborative approach not only enhances efficiency but also promotes resilience within the farming community, enabling the shared use of technology and expertise. The focus moves from individual optimization to wider societal benefits, building sustainable agricultural practices within whole regions.

Integrating sensors within the irrigation system, for instance, allows for real-time monitoring of pressure, flow rates, and even the detection of leaks. This proactive approach reduces downtime and minimizes water loss. Such technologies are cost-effective in the long run by preventing major system failures and ensuring optimal performance. Moreover, data gathered can be analyzed to further refine irrigation scheduling and optimize water usage. Automated system alerts can notify managers of any issues, allowing prompt intervention and avoiding significant problems. This continuous monitoring improves overall system efficiency and reliability, reducing operational costs and enhancing profitability. The use of data analytics helps in the continuous improvement of the whole system, providing valuable insight for optimizing irrigation procedures. Smart data enables smarter decision-making and proactive problem-solving.

7.1 Smart Irrigation Systems

Smart irrigation systems represent a significant advancement in water management for drip irrigation, offering precision and efficiency far beyond traditional methods. These systems leverage technology to optimize water delivery, reducing waste and maximizing crop yields. Imagine a farmer, no longer bound to a rigid schedule, but instead guided by real-time data on soil moisture and weather conditions. This is the power of smart irrigation.

The core of a smart irrigation system lies in its network of sensors. These sensors, strategically placed throughout the field, continuously monitor soil moisture levels, temperature, and even the plant's own health indicators. This data is then transmitted wirelessly to a central control unit, often a computer or a dedicated irrigation controller. This continuous feedback loop allows for incredibly precise adjustments to irrigation scheduling. Unlike traditional systems that rely on fixed schedules, smart systems adapt dynamically to the changing needs of the crops and the environment.

Data analysis is key. The control unit processes the sensor data, using sophisticated algorithms to determine the precise amount of water required for each section of the field. This is a far cry from the blanket irrigation of older methods; smart systems provide site-specific irrigation, catering to the unique needs of individual plants and areas within the field. Factors like sunlight exposure, soil type, and even plant health are all taken into consideration, leading to a more holistic and effective irrigation strategy.

Automation is another key feature of smart irrigation. Once the control unit has processed the data and determined the irrigation needs, it automatically adjusts the flow rates of water to each emitter. This eliminates the need for manual adjustments, freeing up valuable time and labor resources. The system can also be programmed to integrate with weather forecasts, automatically adjusting irrigation schedules in response to predicted rainfall or temperature fluctuations. This proactive approach minimizes water waste and ensures optimal conditions for crop growth.

However, implementing a smart irrigation system requires careful planning and consideration. The initial investment cost can be higher than traditional systems, but the long-term benefits—both environmental and economic—often outweigh the upfront expense. Thorough site assessment is crucial, determining the optimal placement of sensors to ensure accurate data collection. This will differ based on the field layout, crop type, and soil characteristics. Selecting reliable sensors and a robust communication network is equally important; system failures can have major impacts on the entire irrigation process.

Furthermore, proper system maintenance is essential for long-term performance. Regularly inspecting and calibrating sensors and ensuring the reliable operation of the control unit are critical factors in ensuring the continued effectiveness of the investment. Regular cleaning of emitters and pipes should be maintained in sync with the needs of the system. Keeping the system functioning efficiently will deliver a longer return on investment and prevent costly downtimes. Consider creating a comprehensive maintenance schedule and training your staff on how to identify and resolve minor issues. This proactive approach will significantly extend the life span of the system.

In essence, smart irrigation systems represent a paradigm shift in water management for agriculture. They provide a powerful tool for optimizing water use, improving crop yields, and reducing the environmental impact of irrigation. Although the upfront costs may seem high, the long-term economic and environmental benefits make them a compelling investment for both small and large-scale agricultural operations. By embracing these technologies, farmers can ensure a more sustainable and productive future for their operations. The future of irrigation is undeniably smart and precise.

7.2 Precision Agriculture

Precision agriculture leverages technology to optimize resource use and maximize crop yields. It moves beyond blanket irrigation approaches, focusing instead on delivering water and nutrients precisely where and when they are needed most. This targeted approach dramatically improves efficiency, reduces waste, and enhances profitability. Think of it as tailoring your irrigation to each individual plant's specific requirements, rather than treating the entire field uniformly. This is achieved through a combination of advanced sensors, data analysis, and sophisticated irrigation control systems.

Unlike traditional methods, precision agriculture employs sophisticated sensors to monitor various soil and plant parameters. Soil moisture sensors provide real-time data on water content, allowing for precise irrigation scheduling. This eliminates overwatering and underwatering, preventing both water waste and crop stress. Furthermore, sensors can monitor plant health indicators like leaf temperature and chlorophyll content, providing early warnings of potential issues. This proactive approach allows for timely intervention, minimizing losses and maximizing yields.

The data collected by these sensors feeds into sophisticated software programs, enabling the generation of detailed maps of the field. These maps illustrate areas that require more or less water and nutrients. This geospatial analysis allows for variable-rate irrigation, delivering tailored amounts of water to different sections of the field based on their specific needs. This targeted approach drastically reduces water and fertilizer use compared to traditional methods, significantly lowering environmental impact and production costs.

Implementing precision agriculture requires careful planning and investment. Firstly, a thorough site assessment is crucial, requiring an understanding of the field's topography, soil characteristics, and crop requirements. This detailed assessment informs the selection of appropriate sensors and irrigation equipment. Selecting the correct sensors is vital, as inaccurate data compromises the system's effectiveness. Then, establishing a robust data management system is crucial for storing, analyzing, and interpreting sensor data.

Choosing the right technology is critical. While sophisticated systems offer greater precision, simpler solutions might be more appropriate for smaller farms or those with limited budgets. Therefore, a thorough cost-benefit analysis, considering factors such as sensor accuracy, data processing capabilities, and long-term maintenance costs, should be undertaken before investing in any specific technology. This ensures you choose a system that effectively balances your needs and resources, avoiding unnecessary expenses and complexities.

Successful integration of precision agriculture relies heavily on robust data analysis. Interpreting the data generated by sensors and using this information to inform irrigation management decisions is critical. This requires expertise in data analysis and a good understanding of agricultural principles. Farmers may need to develop new skills or seek assistance from agricultural consultants to effectively leverage the data provided by precision agriculture systems. The capacity to utilize and interpret this complex dataset may be crucial.

Beyond water management, precision agriculture facilitates optimized nutrient application. By combining sensor data with soil analysis, farmers can precisely target fertilizer applications, reducing waste and mitigating environmental concerns. This targeted fertilization also improves nutrient use efficiency, resulting in healthier plants and higher yields.

It's essential to recognize that this strategy, while beneficial, requires a more in-depth understanding of crop nutrition.

Maintaining a precision agriculture system requires regular calibration and maintenance of sensors and irrigation equipment. This ensures the system continues to accurately monitor field conditions and deliver water and nutrients efficiently. Regular maintenance also helps prevent equipment failure and extends the lifespan of the system. Ignoring maintenance leads to poor data, impacting irrigation decisions and causing production losses. This aspect is often overlooked but fundamental for successful implementation.

Precision agriculture isn't just about technology; it demands a shift in mindset and farming practices. It requires farmers to adopt a more data-driven approach to decision-making, relying on objective information rather than intuition or historical practices. This shift requires embracing new skills and technologies, sometimes involving significant investment in training and equipment. However, the resulting improvements in efficiency and yield make this investment worthwhile in the long run.

Finally, precision agriculture is constantly evolving. New technologies and techniques are continually being developed, offering improved accuracy, efficiency, and functionalities. Staying informed about these advancements is essential for maintaining a competitive edge and optimizing farm operations. Attending workshops, conferences, or online courses helps farmers adapt to the latest developments and remain at the forefront of this revolutionary agricultural practice.

7.3 Case Studies

A vineyard in the sun-drenched hills of Tuscany faced a persistent challenge: inconsistent water distribution across its sloping terrain. Traditional furrow irrigation led to waterlogging in low-lying areas and insufficient moisture in higher regions, resulting in uneven grape ripening and reduced yield. Implementing a drip irrigation system, however, transformed their operation. A carefully designed network of micro-tubes, strategically placed to account for the slope, delivered water directly to the base of each vine. This precise application not only conserved water—reducing their consumption by 40%—but also significantly improved grape quality and increased their harvest by 15%. The consistent moisture levels promoted even ripening, resulting in superior wine quality and a higher market price. This exemplifies the transformative power of well-planned drip irrigation in optimizing water use and improving product quality even in challenging geographical conditions.

Consider a large-scale commercial tomato farm in arid Arizona. Facing escalating water costs and the threat of drought, the farm adopted a comprehensive drip irrigation strategy. The initial investment was substantial, encompassing specialized soil sensors, automated control systems, and a sophisticated water filtration system. Yet, the long-term returns have justified the expense. Real-time soil moisture monitoring provided precise data to optimize irrigation scheduling, minimizing water waste and maximizing yield. Furthermore, the controlled delivery of nutrients via fertigation enhanced plant growth and fruit quality, leading to increased profits. Data analysis revealed a remarkable 30% reduction in water usage compared to their previous flood irrigation practices, while simultaneously achieving a 20% increase in tomato yield. This project underscores the economic viability of advanced drip irrigation solutions even in harsh environments.

A small-scale organic farm in rural Kenya initially struggled with inconsistent rainfall and limited access to water resources. The adoption of a simple, low-cost drip irrigation system, however, proved to be a game-changer. Using locally sourced materials and employing community-based training programs, the farmers successfully implemented a system tailored to their specific needs and resources. The system relied on gravity-fed water from a nearby spring, minimizing energy costs. The precise delivery of water directly to the roots of their crops drastically reduced water waste and enabled efficient cultivation during extended dry periods. This resulted in increased crop yields and improved food security for the community, showcasing the adaptability of drip irrigation in resource-constrained settings and its significant socio-economic impact.

In a contrasting scenario, a hydroponic lettuce farm in a controlled environment utilized a high-tech drip irrigation system integrated with advanced sensors and automated control mechanisms. This system monitored environmental factors like temperature and humidity, adjusting water delivery accordingly for optimal lettuce growth. The precision of water and nutrient delivery significantly minimized waste and ensured consistent crop quality. The use of recirculating water systems further reduced water consumption, minimizing environmental impact and contributing to sustainable agricultural practices. The automated system significantly reduced labor costs associated with irrigation management, improving efficiency and profitability for the operation. This high-tech approach demonstrates the scalability and adaptability of drip irrigation to various cultivation methods and environments, highlighting its economic and environmental benefits.

Finally, an olive grove in Spain facing the challenges of both water scarcity and soil erosion significantly improved its productivity and sustainability through a cleverly designed drip irrigation system. The system incorporated techniques to minimize soil erosion, such as using mulch and strategically placed drip lines. Furthermore, the carefully planned irrigation schedule, guided by soil moisture sensors, ensured optimal water utilization while conserving precious water resources. The result was a substantial increase in olive yield and a marked improvement in the overall health and resilience of the olive trees. The case study illustrates the versatility of drip irrigation in addressing multiple agricultural challenges and promoting environmental stewardship.

www.ingramcontent.com/pod-product-compliance
Lightning Source LLC
LaVergne TN
LVHW022014200926
848481LV00007B/2387